PICTOGRAPHS

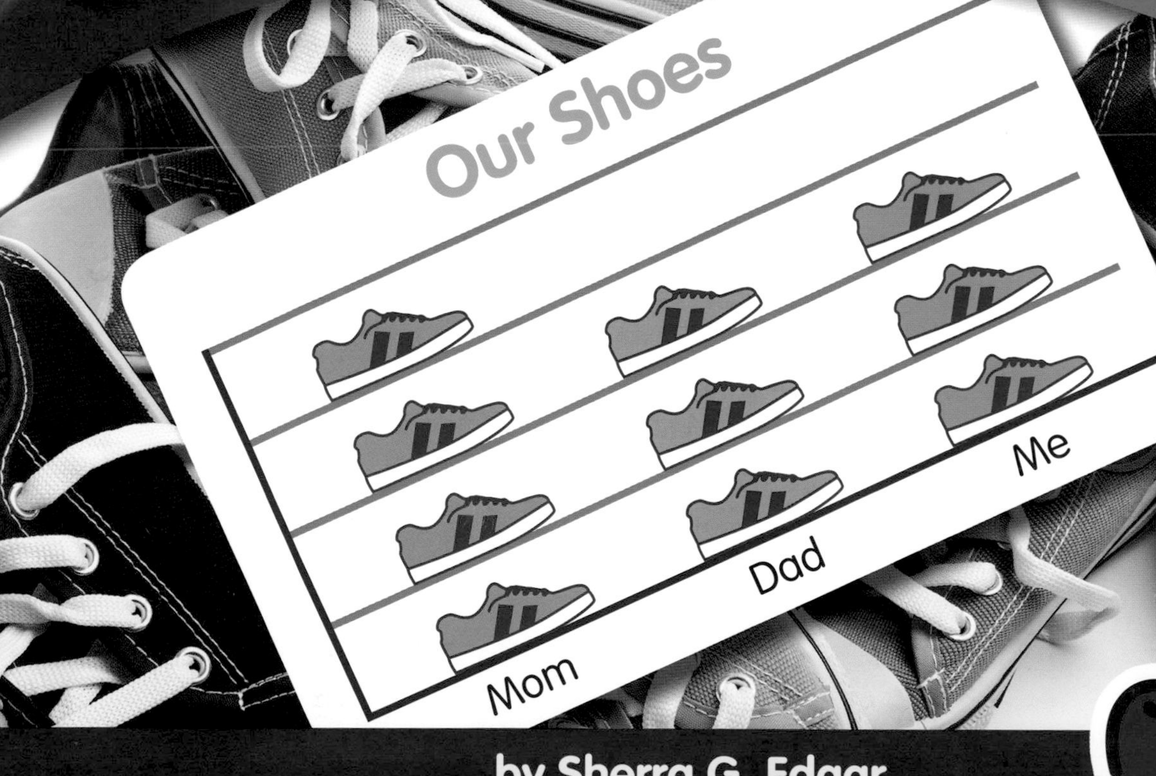

by Sherra G. Edgar

Cherry Lake Publishing • Ann Arbor, Michigan

2

Published in the United States of America by Cherry Lake Publishing
Ann Arbor, Michigan
www.cherrylakepublishing.com

Consultants: Janice Bradley, PhD, Mathematically Connected Communities, New Mexico State University; Marla Conn, Read-Ability
Editorial direction and book production: Red Line Editorial

Photo Credits: Jiri Hera/Shutterstock Images, cover, 1; Jerry Sharp/Shutterstock Images, 4; Shutterstock Images, 6, 8, 16, 20; Richard Lister/Shutterstock Images, 10; Thomas M Perkins/Shutterstock Images, 12; Pavel L Photo and Video/Shutterstock Images, 18

Library of Congress Cataloging-in-Publication Data
Edgar, Sherra G.
Pictographs / Sherra G. Edgar.
pages cm. -- (Let's make graphs)
Audience: K to grade 3.
Includes bibliographical references and index.
ISBN 978-1-62431-393-6 (hardcover) -- ISBN 978-1-62431-469-8 (paperback) -- ISBN 978-1-62431-431-5 (pdf) -- ISBN 978-1-62431-507-7 (ebook)
1. Graphic methods--Juvenile literature. 2. Picture-writing--Juvenile literature. 3. Signs and symbols--Juvenile literature. 4. Charts, diagrams, etc.--Juvenile literature. I. Title.

QA90.E35 2013
518'.23--dc23

2013004938

Cherry Lake Publishing would like to acknowledge the work of The Partnership for 21st Century Skills. Please visit *www.p21.org* for more information.

Printed in the United States of America
Corporate Graphics Inc.
July 2013
CLFA11

TABLE OF CONTENTS

What Is a Pictograph?

Chris wants to count what color fish he sees. He can use a **graph**. Graphs show **data**.

Fish

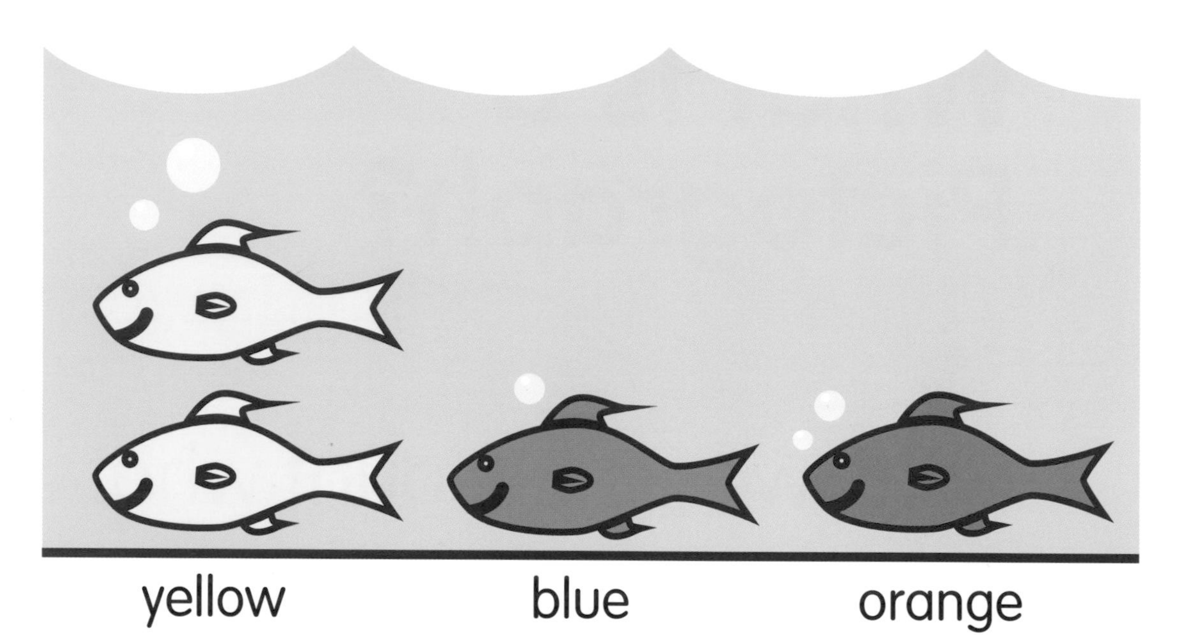

Chris made a **pictograph**. A pictograph is a graph that uses pictures. The pictures stand for **amounts**. They are **symbols**.

Favorite Kinds of Ice Cream

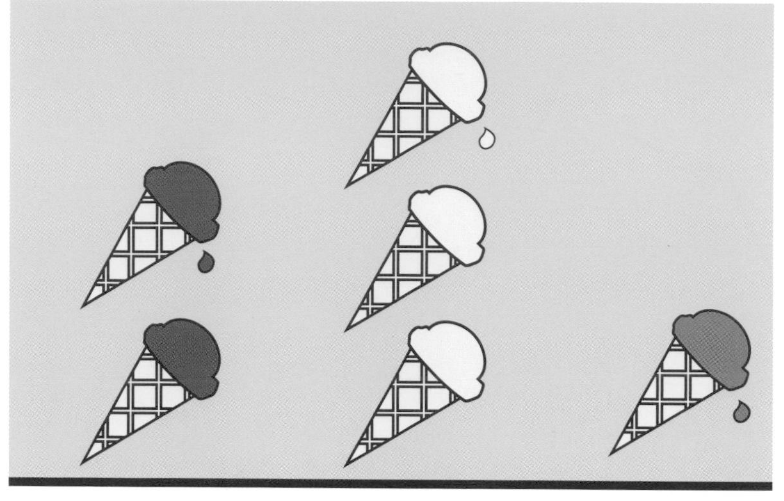

A symbol can stand for one or many things. The **key** tells how many things each symbol stands for.

Duck Colors

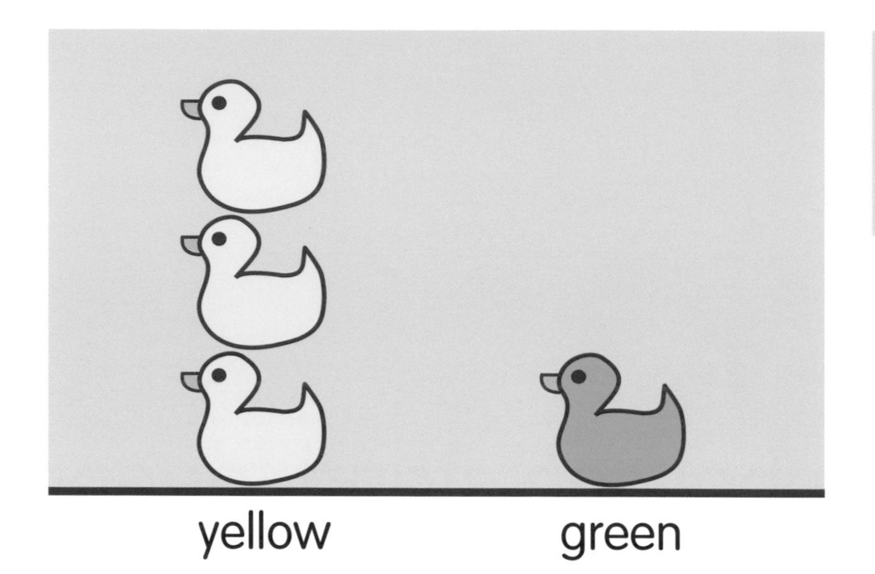

More symbols mean bigger amounts. Fewer symbols mean smaller amounts. Symbols help us **compare** data.

Making a Pictograph

Mia is counting her friends' teddy bears. She calls her graph "Teddy Bears."

Teddy Bears

Mia	Rosa	Zach

Mia drew a line. She wrote everyone's names below. This shows whose bears she counted.

Teddy Bears

Mia used a bear face as the symbol. Each bear face stands for two teddy bears.

Teddy Bears

Key

 = two teddy bears

How many teddies does Zach have? Count two for each bear face. Zach has eight teddies.

You Try It!

Make a pictograph with fruit. Use pictures to stand for kinds of fruit. Graph the fruit you eat!

Find Out More

BOOK

Leedy, Loreen. *The Great Graph Contest*. New York: Holiday House, 2006.

WEB SITE

IXL Math: Interpreting Graphs

http://www.ixl.com/math/kindergarten/interpreting-graphs

Practice reading pictographs at this web site.

Glossary

amounts (uh-MOUNTS) how many or how much there is of something

compare (kuhm-PAIR) to show how things are alike

data (DEY-tah) amounts from a graph

graph (GRAF) a picture that compares two or more amounts

key (KEE) a table that tells about the symbols on a pictograph

pictograph (PIK-tuh-graf) a graph that uses pictures to show amounts

symbols (SIM-buhls) things that stand for something else

Home and School Connection

Use this list of words from the book to help your child become a better reader. Word games and writing activities can help beginning readers reinforce literacy skills.

a
amounts
are
as
bears
below
bigger
blue
calls
can
chocolate
Chris
color
compare
count
counted
counting
cream
data
does
drew
duck
each
eat
eight
everyone
face
favorite
fewer
fish
for
friends
fruit
graph
graphs
green
have
he
help
her
how
ice
is
it
key
kinds
line
made
make
making
many
mean
Mia
more
names
of
one
or
orange
people
pictograph
pictures
Rosa
sees
she
show
shows
smaller
stand
stands
strawberry
symbol
symbols
teddies
teddy
tells
that
the
they
things
this
to
try
two
us
use
used
uses
vanilla
wants
what
whose
wrote
yellow
you
Zach

Index

About the Author

Sherra G. Edgar is a former primary school teacher who now writes books for children. She also writes a blog for women. She lives in Texas with her husband and son. She loves reading, writing, and spending time with friends and family.